DIGITAL

AN EXPOSITORY GUIDE ON HOW TO MAKE USE OF THE DIGITAL MULTIMETER

ELIANA FRED

copyright@2021

Table of Contents

CHAPTER ONE 3
   INTRODUCTION 3
CHAPTER TWO 6
   HOW TO SELECT THE BEST MULTIMETER 6
CHAPTER THREE 21
   IMPORTANT MAINTENANCE TIPS FOR DIGITAL MULTIMETERS 21
CHAPTER FOUR 24
   MAKING USE A MULTIMETER TO MEASURE VOLTAGE, CURRENT AND RESISTANCE 24
CHAPTER FIVE 29
   COMPONENT OF A METER 29
CHAPTER SIX 45
   MEASURING WATTAGE AND THE POWER CONSUMPTION OF AN APPLIANCE WITH A MULTIMETER 45
CHAPTER SEVEN 50
   ANSWERS TO FREQUENTLY ASKED QUESTIONS ABOUT THE MULTIMETERS 50
CHAPTER EIGHT 62
   DIGITAL MULTIMETER USES AND FUNCTIONS 62
THE END 65

CHAPTER ONE

INTRODUCTION

What is a multimeter?

A multimeter is a measurement apparatus totally vital in electronics. It consolidates three fundamental highlights: a voltmeter, ohmeter, and ammeter, and sometimes coherence.

A computerized multimeter or DMM is a helpful test instrument for measuring voltage, current and obstruction, and a few meters have an office for testing transistors and capacitors. You can likewise utilize it for checking congruity of wires and circuits. In the event that you like to DIY, do vehicle upkeep or investigate electronic or electrical hardware, a multimeter is a helpful assistant to have in your home toolbox.

A multimeter permits you to comprehend what is happening in your circuits. At whatever point something in your circuit isn't working.

Here are some situations in electronics projects that you will find the multimeter very useful:

- is the switch on?
- is this wire directing the power or is it broken?
- how much current is moving through this driven?
- how much force do you have left on your batteries?

These and different inquiries can be replied with the assistance of a multimeter.

CHAPTER TWO

HOW TO SELECT THE BEST MULTIMETER

You can locate a wide assortment of multimeters with various functionalities and precision. A fundamental multimeter costs about $5 and measures the three easiest yet most significant qualities in your circuit: voltage, current, and obstruction.

Anyway you can figure that this multimeter won't last more and isn't exact. The best multimeter for you will rely upon what you plan to do, in case you're an apprentice or an expert circuit repairman, and on your financial plan.

Getting acquainted with a multimeter

A multimeter is formed by four fundamental segments:

- Display: this is the place the estimations are shown
- Selection handle: this chooses what you need to gauge
- Ports: this is the place you plug in the tests

- Probes: a multimeter accompanies two tests. By and large one is red and the other is dark.

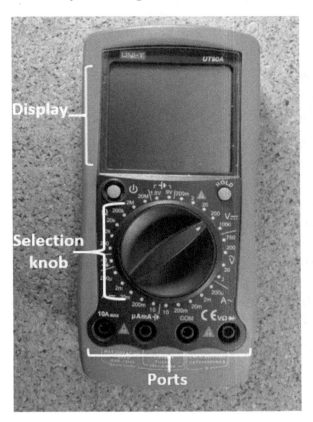

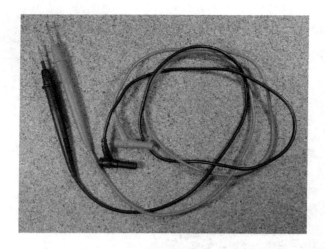

Note: There isn't any contrast between the red and the dark tests, only the shading.

Along these lines, accepting the show:

- the dark test is constantly associated with the COM port.
- the red test is associated with one of different ports relying upon what you need to gauge.

Ports

The "COM" or "– " port is the place the dark test ought to be associated. The COM test is customarily dark.

- 10A is utilized when estimating enormous currents, more noteworthy than 200mA
- µAmA is utilized to quantify current
- VΩ permits you to quantify voltage and obstruction and test coherence

This ports can change contingent upon the multimeter you're utilizing.

Measuring Voltage

You can gauge DC voltage or AC voltage. The V with a straight line implies DC voltage.

The V with the wavy line implies AC voltage.

To gauge voltage:

1. Set the mode to V with a wavy line in case you're estimating AC voltage or to the V with a straight line in case you're estimating DC voltage.

2. Make sure the red test is associated with the port with a V close to it.

3.   Connect the red test to the positive side of your part, which is where the current is coming from.

4.   Connect the COM test to the opposite side of your segment.

5.   Read the incentive on the showcase.

Tip: to quantify voltage you need to associate your multimeter in corresponding with the part you need to gauge the voltage. Setting the multimeter in equal is putting each test along the leads of the segment you need to quantify the voltage.

Example: estimating a battery's voltage

In this model we're going to quantify the voltage of a 1.5V battery. You realize that you'll have around 1.5V. Along these lines, you should choose a range with the determination handle that can peruse the 1.5V. So you should choose 2V on account of this multimeter. In the event that you have an autorange multimeter, you don't need to stress over the range you have to choose.

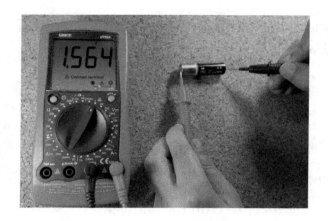

Consider the possibility that you didn't have the foggiest idea what was the estimation of the voltage. On the off chance that you have to quantify the voltage of something, and you don't have the foggiest idea about the

range wherein the worth will fall under, you have to attempt a few extents.

On the off chance that the range you've chosen is lower than the genuine worth, on the showcase you'll peruse 1 as appeared in the image beneath. The 1 implies that the voltage is higher than the range you've chosen.

In the event that you select a higher range, most piece of the occasions you'll have the option to peruse the estimation of the voltage, yet with less exactness.

What occurs on the off chance that you switch the red and the dark test?

Nothing perilous will occur. The perusing on the multimeter has a similar worth, however it's negative.

Model: estimating voltage in a circuit

In this model we'll tell you the best way to gauge the voltage drop over a resistor in a basic circuit. This model circuit illuminates a LED.

TIP: two parts in equal offer voltage, so you ought to interface the multimeter tests in corresponding with the segment you need to gauge the voltage.

To wire the circuit you have to interface a LED to 9V battery through a 470 Ohm resistor.

To gauge the voltage drop over the resistor:

1. You simply need to put the red test in one lead of the resistor and the dark test on the other lead of the resistor.

2. The red test ought to be associated with the part that the current is originating from.

3. Also, remember to ensure the tests are connected the correct ports.

Measuring Current

To quantify current you have to manage at the top of the priority list that segments in arrangement share a current. In this way, you have to associate your multimeter in arrangement with your circuit.

TIP: to put the multimeter in arrangement, you have to put the red test on the lead of a segment and the dark test on the following segment lead. The multimeter goes about as though it was a wire in your circuit. On the off chance that you disengage the multimeter, your circuit won't work.

Prior to estimating the current, be certain that you've connected the red test in the correct port, for this situation μAmA. In the model beneath, a similar circuit of the past model is utilized. The multimeter is a piece of the circuit.

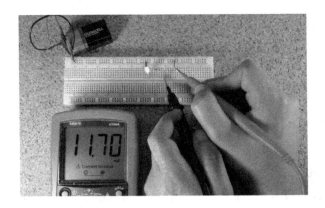

Measuring Resistance

Fitting the red test into the correct port and turn the determination handle to the opposition segment. At that point, associate the tests to the resistor leads. The manner in which you associate the leads doesn't make a difference, the outcome is the equivalent.

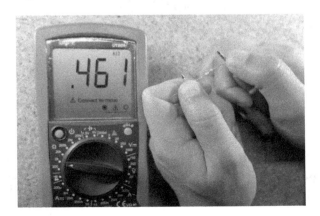

As should be obvious, the 470ω resistor, just has 461ω.

Checking Continuity

Most multimeters give a component that permits you to test the coherence of your circuit. This permits you to effortlessly identify bugs, for example, broken wires. It likewise causes you check if two purposes of the circuit are associated.

To utilize this usefulness select the mode that resemble a speaker.

How does continuity work?

In the event that there is extremely low obstruction between two focuses, which is not exactly a couple of ohms, the two focuses are electrically associated and you'll hear a persistent sound.

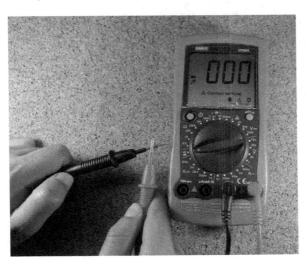

On the off chance that the sound isn't persistent or on the off chance that you don't hear any solid whatsoever, It implies that what you're trying has a broken association or isn't associated in any way.

Cautioning: To test progression you should kill the framework! Mood killer the force gracefully!

Contact the two tests together and, as they are associated, you'll hear a consistent sound.

To test the congruity of a wire, you simply need to interface each test to the wire tips.

Wrapping up

A multimeter is a fundamental device in any hardware lab. In this current Beginner's Guide, we've told you The best way To Use a Multimeter. You've figured out how to quantify voltage, current and opposition, and how to check congruity.

CHAPTER THREE

IMPORTANT MAINTENANCE TIPS FOR DIGITAL MULTIMETERS

The more exact your electrical estimating instrument is, the better you are as a professional. This is the basic truth that regularly goes disregarded by numerous specialists. Consequently, makers of top notch items wind up purchasing exactly adjusted test hardware. Keeping up the hardware, for example, a digital multimeter is of most extreme significance, to guarantee precise perusing of voltage and generally electrical productivity. So here are a couple of significant hints to keeping up your multimeter.

1. The right setting

Ensure that your multimeter is on the right setting for the current test. For instance, if your multimeter is set at 20VDC scale and you're estimating 120VAC, the screen will show "OL".

While this doesn't really hurt your meter, it might influence its precision.

2.Cleaning

No profound cleaning is needed for multimeters. You can do it by quickly cleaning it with a clammy (not wet) fabric on its surface. You can undoubtedly discover particular cleaning wipes for multimeters to get exact readings and to keep them dust free.

3.Storage

Ensure that you store your digital multimeter just in a dry spot, to keep away from them getting harmed because of dampness or water. Additionally, putting away them for a situation will guarantee assurance from actual harm. Likewise, in case you're anticipating putting away it for an extended length of time, eliminate its batteries. This will help hold erosion back from gathering at the associations.

Your estimating and test gear should be maneuvered carefully to keep away from any setbacks or erroneous readings. In the event that need to extend your Quality Control office and are searching for an organization that sell and adjusts test hardware, connect with us at Industrial Instrument Works. We are approved merchants for more than 100 driving makers of test and estimation instrumentation

CHAPTER FOUR

MAKING USE A MULTIMETER TO MEASURE VOLTAGE, CURRENT AND RESISTANCE

Volts, Amps, Ohms - What Does everything Mean?

Before we figure out how to utilize a multimeter, we have to get comfortable with the amounts we will be estimating. The most fundamental circuit we will experience is a voltage source, which could be associated with a heap. The voltage source could be a battery or a mains power gracefully. The heap may be an apparatus, for example, a bulb or electronic segment called a resistor. The circuit can be spoken to by a chart called a schematic. In the circuit beneath, the voltage source V makes an electrical weight which powers a current I to stream around the circuit and through the heap R. Ohm's Law reveals to us that on the off chance that we separate the voltage V by the opposition R, estimated in ohms, it gives us an incentive for the current I in amps:

$V/R = I$

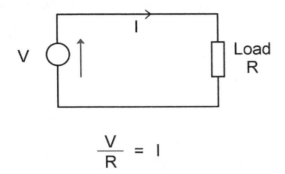

$$\frac{V}{R} = I$$

A straightforward circuit comprising of an AA cell and a bulb. The AA cell is the voltage source that makes current stream around the circuit and through the bulb. | Source

Amounts and Terms Used in Electrical Engineering

Volts

This is the weight between two focuses in an electrical circuit. It could be estimated over the voltage source or different segments associated in the circuit.

Amps

This is a proportion of the current streaming between two focuses in an electrical circuit.

Ohms

A proportion of the protection from stream in a circuit.

Voltage Source

This delivers a current stream in a circuit. It could be a battery, compact generator, mains flexibly to a home, alternator on your motor or seat power gracefully in a lab or workshop.

Burden

A gadget or part which draws power from a voltage source. This could be an electronic resistor, bulb, electric warmer, engine or any electrical apparatus. A heap has an opposition estimated in ohms.

Ground

This is generally the point in a circuit to which the negative terminal of a battery or force flexibly is associated.

DC

Direct current. Current streams just a single route from a DC source, a case of which is a battery.

AC

Alternating Current. Current streams one route from a source, inverts, and afterward streams the other way. This happens all the time at a rate controlled by the recurrence which is normally 50 or 60 hertz. The mains gracefully in a house is AC.

Extremity

A term used to portray the course of stream of current in a circuit or which focuses are certain and which are negative wrt a reference point.

For increasingly point by point data about these amounts and terms, take a temporary re-route to my other article:

What Does a Multimeter Measure?

A fundamental multimeter encourages the estimation of the accompanying amounts:

- DC voltage
- DC current

- AC voltage

- AC current (not every single essential meter have this capacity)

- Resistance

- Continuity - showed by a bell or tone

Likewise meters may have the accompanying capacities:

- Capacitance estimation

- Transistor HFE or DC current increase

- Temperature estimation with an extra test

- Diode test

- Frequency estimation

The worth estimated by the instrument is shown on a LCD show or scale.

CHAPTER FIVE

COMPONENT OF A METER

• The Display. This is typically a multidigit, 7 portion LCD show. Some research center instruments anyway have LED shows which are simpler to peruse under certain lighting conditions.

• Rotary Range Selector Dial. This permits you to choose the capacity which you will use on the meter. On a non-autoranging meter, it likewise chooses the range.

• Connection Sockets. These are 4mm breadth female attachments into which 4 mm test leads are stopped. The course of action is non-standard and relies upon the brand/model of meter, so it's imperative to comprehend the capacity of every attachment to keep away from harm to the meter:

Com is the basic attachment into which the negative or ground lead is stopped.

On the off chance that an attachment is checked VΩmA, this is the attachment into which the positive test lead is stopped for estimating voltage, opposition or current ("mA" signifies "milliamps"). On the off chance that there is no notice of "An" or "Mama" on this attachment, there will be at least one separate attachments for interfacing the test lead to quantify current. These extra attachments might be denoted "An" or "Mama" with the current rating (for example 10A for high current readings and 400 mA for lower current readings).

How Do I Setup a Multimeter to Measure Volts, Amps or Ohms?

Voltage, current and obstruction ranges are typically set by turning a rotating range choice dial. This is set to the amount being estimated, for example Air conditioning volts, DC volts, Amps(current) or Ohms (obstruction).

On the off chance that the meter is non-autoranging, each capacity will have a few extents. So for instance, the DC volts work range will have 1000V, 200V, 20V, 2V and 200mV territories. Utilizing the most minimal range potential gives progressively critical figures in the perusing.

Uncovered conductor of test lead | Source

Step by step instructions to Measure Voltage

1. Power off the circuity/wiring under test if there is a threat of shorting out firmly divided nearby wires, terminals or different focuses which have varying voltages.

2. Plug the dark ground test lead into the COM attachment on the meter (see photograph beneath).

3. Plug the red positive test lead into the attachment stamped V (typically likewise set apart with the Greek letter "omega" Ω and perhaps a diode image).

4. If the meter has a manual range choice dial, go this to choose AC or DC volts and pick a range to give the necessary precision. So for example estimating 12 volts on the 20 volt range will give more decimal spots than on the 200 volt go.

In the event that the meter is autoranging, turn the dial to the 'V' setting with the image for AC or DC (see "What Do the Symbols on the Range Dial Mean?" beneath).

5. A multimeter must be associated in equal in a circuit (see graph underneath) so as to quantify voltage. So this implies the two test tests ought to be associated in corresponding with the voltage

source, load or some other two focuses across which voltage should be estimated.

6. Touch the dark test against the main purpose of the hardware/wiring.

7. Power up the hardware.

8. Touch the other red test against the second purpose of test. Guarantee you don't overcome any barrier between the fact of the matter being tried and contiguous wiring, terminals or tracks on a PCB.

9. Take the perusing on the LCD show.

Note: A lead with a 4mm banana plug toward one side and a crocodile cut on the opposite end is exceptionally helpful. The croc clasp can be associated with ground in the circuit, opening up one of your hands.

Associating Probe Leads to Measure Voltage

Test leads and 4mm attachments on a DMM, arrangement to quantify voltage | Source

Arrangement and Parallel Connections

Clarifying arrangement and equal associations (R1, R2 and R3 are resistors) | Source

Estimating Voltage - Meter in Parallel With Load or Voltage Source

Caution

When estimating mains voltages, consistently turn off force before associating estimating tests. In the event that this is beyond the realm of imagination, consistently interface with the unbiased first.

Security First When Measuring Mains Voltages!

1. Before utilizing a meter to quantify mains voltages, guarantee test leads aren't harmed and that there are no uncovered conductors which could be contacted coincidentally.

2. Double watch that test leads are connected to the normal and voltage attachments of the DMM (see photograph beneath) and not the current attachments. This is basic to abstain from exploding the meter.

3. Set the range dial on the meter to AC volts and the most noteworthy voltage run.

4. If you need to check the voltage at an attachment outlet, switch off force utilizing the switch on the attachment. At that point embed tests into the mains attachment. In the event that the attachment outlet has no switch and you can't kill power, embed a test into the unbiased pin first before embeddings a test into the hot (live) pin of the attachment. In the event that you embed the test into the hot (live) pin first and the meter is flawed, current could course through the meter to the unbiased test. On the off chance that you, at that point accidentally contact the tip of the test or the test is left on a conductive metal surface, there is a chance of stun.

5. Probes with crocodile cuts permit associations with be made with power killed and don't need to be held set up when force is turned on.

6. Finally turn on the force switch and measure the voltage.

In a perfect world purchase and utilize a meter with a least CAT III or ideally CAT IV security for testing mains voltages. This kind of meter will fuse high bursting limit (HRC) wires and other inner wellbeing parts that offer the most elevated level of security against over-burdens and homeless people on the line being tried. A meter with less assurance can conceivably explode causing injury on the off chance that it is associated mistakenly, or a transient voltage creates an inner circular segment.

Autoranging Meters

Autoranging meters recognize the greatness of the voltage and select the range consequently to give the most measure of huge digits on the presentation. You should anyway set the mode to opposition, volts or current and furthermore interface the test prompts the best possible attachments when estimating current.

What Do the Symbols on the Range Dial Mean?

Recognizing Live or Hot Wires

A Fluke "VoltAlert™" non-contact voltage indicator is a standard device in any circuit repairmen toolbox, however helpful for property holders moreover. I utilize one of these for distinguishing which conductor is live at whatever point I'm doing any home upkeep. Not at all like a neon screwdriver analyzer (stage analyzer), you can utilize one of these in circumstances when live parts/wires are covered or secured with protection and you can't reach wires. It additionally proves to be handy for checking whether there's a break in a force flex and where the break happens.

Note: It's consistently a smart thought to utilize a neon analyzer to twofold watch that force is unquestionably off while doing any electrical upkeep.

Instructions to Measure Current

1.   Turn off the force in the circuit being estimated.

2.   Connect the test leads as appeared in the photograph underneath. Fitting the dark ground test lead into the COM attachment.

3.   Plug the red positive test lead either into the mA attachment or the high current attachment which is normally stamped 10A (a few meters have a 20 An attachment rather than 10A). The mA attachment is regularly set apart with the most

extreme current and in the event that you gauge that the current will be more prominent than this worth, you should utilize the 10 An attachment, else you will wind up blowing a wire in the meter. On certain meters, there is no extra attachment for estimating current and a similar attachment is utilized with respect to estimating voltage (normally stamped VΩmA).

4. A multimeter must be embedded in arrangement in a circuit so as to gauge current. See the graph beneath.

5. Turn the dial on the meter to the most elevated current range (or the 10A territory if the test is in the 10A attachment). In the event that the meter is autoranging, set it to the "An" or mA setting. (See the photograph above for a clarification of images utilized).

6. Turn on the force.

7. If the range is excessively high, you can change to a lower range to get a progressively precise perusing.

8. Remember to restore the positive test to the V attachment when got done with estimating

current. The meter is essentially a short out when the lead is in the mA or 10 An attachment. On the off chance that you overlook and associate the meter to a voltage source when the lead is in this position, you may wind up blowing a breaker, best case scenario or exploding the meter best case scenario! (On certain meters the 10A territory is un-combined).

Interfacing Probe Leads

Interfacing Probe Leads to Measure Current

Estimating Current - Meter in Series

What Multimeter Should I Buy?

When asked, Fluke, who are a main US producer of advanced instrumentation, suggested the Fluke 113 model for broadly useful use in the home or for vehicle support. This is a superb meter and can gauge AC and DC volts, opposition, check congruity and diodes. The meter is auto-extending, so runs don't need to be set. It is likewise a genuine RMS meter. It doesn't gauge current, so If you have to quantify AC and DC current, the Fluke 115 has this additional office.

An option is the Fluke 177 model which is a high exactness instrument (the detail is 0.09% precision on DC volts). I utilize this model for progressively precise testing and expert use and it can gauge AC and DC voltage and current, obstruction, recurrence, capacitance, coherence and diode test. It can likewise demonstrate max and min esteems on each range.

Accident 177 Multimeter with Auto-Ranging Facility

Estimating Large Currents with a Clamp Meter (Tong Tester)

On most multimeters, the most elevated current range is 10 or 20 amps. It is unrealistic to take care of high flows through a meter since ordinary 4 mm attachments and test leads wouldn't be equipped for conveying high flows without overheating. Rather, cinch meters are utilized for these estimations.

Brace meters (as the name proposes), otherwise called tong analyzers, have a spring stacked cinch like a mammoth garments peg which clasps around a current conveying link. The upside of this is a circuit doesn't need to broken to embed a meter in arrangement, and force needn't be killed just like the situation when estimating current on a standard DMM. Clasp meters utilize either a coordinated current transformer or corridor impact sensor to quantify the attractive field created by a streaming current. The meter can be an independent instrument with a LCD which shows current, or on the other hand the gadget can yield a voltage signal by means of test leads and 4mm "banana" attachments to a standard DMM. The voltage is relative to the deliberate sign, normally 1mv speaks to 1 amp.

Clasp meters can quantify hundreds or thousands of amps.

To utilize a current clasp, you just clip over a solitary link. On account of a force string or multicore link, you have to detach one of the centers. In the event that two centers conveying a similar current yet in inverse ways are encased inside the jaws (which would be the circumstance on the off chance that you clip over a force string), the attractive fields because of the current stream would counterbalance and the perusing would be zero.

Instructions to Measure Resistance

1. If the segment is on a circuit board or in an apparatus, turn off the force

2. Disconnect one finish of the segment if it's in a circuit. This may include pulling off spade leads or desoldering. This is significant as there might be different resistors or different segments having obstruction, in corresponding with the part being estimated.

3. Connect the tests as appeared in the photograph beneath.

4. Turn the dial to the most minimal Ohm or Ω run. This is probably going to be the 200 ohm run or comparative.

5. Place a test tip at each finish of the part being estimated.

\6. If the presentation designates "1", this implies opposition is more noteworthy than can be shown on the range setting you have chosen, so you should turn the dial to the following most elevated range. Rehash this until a worth is shown on the LCD.

Interfacing Probe Leads to Measure Resistance

The most effective method to Check Continuity and Fuses

A multimeter is helpful for checking breaks in flexes of apparatuses, blown fibers in bulbs and blown wires, and following ways/tracks on PCBs

1. Turn the choosing dial on the meter to the coherence extend. This is regularly demonstrated by an image which appears as though a progression of circular segments of a circle (See the photograph indicating images utilized on meters above).

2. Connect the test prompts the meter as appeared in the photograph beneath.

3. If a channel on a circuit board/a wire in an apparatus should be checked, ensure the gadget is shut down.

4. Place the tip of a test at each finish of the conductor or circuit which should be checked.

5. If obstruction is not exactly around 30 ohms, the meter will show this by a signal tone or

humming sound. The obstruction is typically demonstrated on the presentation too. On the off chance that there is break in progression in the gadget being tried, an over-burden sign, for the most part the digit "1", will be shown on the meter.

Associating Probe Leads to Check Diodes or Continuity

Step by step instructions to Check Diodes

A multimeter can be utilized to check whether a diode is shortcircuited or open circuited. A diode is an electronic one way valve or check valve, which just leads a single way. A multimeter when associated with a working diode demonstrates the voltage over the segment.

1. Turn the dial of the meter to the diode test setting, which is shown by a triangle with a bar toward the end (see the photograph demonstrating images utilized on meters above).

2. Connect the tests as appeared previously.

3. Touch the tip of the negative test to one finish of the diode, and the tip of the positive test to the opposite end.

4. When the dark test is in contact with the cathode of the diode (as a rule demonstrated by a bar set apart on the segment) and the red test reaches the anode, the diode conducts, and the meter shows the voltage. This ought to be about 0.6 volts for a silicon diode and about 0.2 volts for a Schottky diode. At the point when the tests are switched, the meter ought to show a "1" in light of the fact that the diode is open circuit and non-directing.

5. If the meter peruses "1" when the tests are put in any case, the diode is probably going to be defective and open circuit. On the off chance that the meter shows a worth near zero, the diode is shorted circuited.

6. If a part is in circuit, protections in equal will influence the perusing and the meter may not specify "1" however a worth fairly less.

CHAPTER SIX

MEASURING WATTAGE AND THE POWER CONSUMPTION OF AN APPLIANCE WITH A MULTIMETER

Watts = Volts x Current

So to quantify the force in watts of a heap/apparatus, both the voltage over the heap and the current going through it must be estimated. On the off chance that you have two DMMs, you can gauge the voltage and current at the same time. On the other hand measure the voltage first, and afterward disengage the heap with the goal that the DMM can be embedded in arrangement to quantify current. At the point when any amount is estimated, the estimating gadget has an effect on the estimation. So the opposition of the meter will decrease current marginally, and give a lower perusing than the real incentive with the meter not associated.

The most secure approach to gauge the force utilization of an apparatus fueled from the mains is to utilize a force connector. These gadgets plug into an attachment and the apparatus is then connected to the connector which shows data on a LCD. Run of the mill boundaries showed are voltage, current, power, kwh, cost and to what extent the apparatus was turned on (helpful for coolers, coolers and forced air systems which cut in and out). You can peruse progressively about these device in my article here:

Checking Power Consumption of Appliances With an Energy Monitoring Adapter

An elective method of securely estimating flow drawn by an electrical machine is to make up a test lead utilizing a short bit of intensity line with a trailing attachment toward one side and a mains plug on the other. The internal impartial center of the force rope could be liberated and isolated from the external sheath, and current estimated with a cinch meter or test (Don't evacuate the protection!) . Another path is to cut the impartial center, add 4mm banana fittings to every one of the cut closures and attachment these into the meter. Just make associations and alter go on the meter with the force off!

Instructions to Check Peak Voltages - Using a DVA Adapter

A few meters have a catch which sets the meter to understand max and min RMS voltages or potentially top voltages (of the waveform). An option is to utilize a DVA or Direct Voltage Adapter. A few parts, for example, CDI (Capacitor Discharge Ignition) modules on vehicles, vessels and little motors produce beats which differ in recurrence and can be brief span. A DVA connector will test and hold the pinnacle estimation of the waveform and yield it as a DC voltage so the segment can be verified whether it's delivering the right voltage level. A DVA connector regularly has two test leads as contribution for estimating voltage and either two yield leads with banana plugs or a connector with fixed plugs appended for connecting to a meter with standard divided attachments. The meter is set to a high DC voltage extend (for example 1000 volts DC) and the connector regularly yields 1 volt DC for every 1 volt AC input.

Significant data for anybody utilizing a DVA to check start circuits!

In this application, the connector is utilized for estimating the essential voltage of a stator/start loop, not the optional voltage, which could be around 10,000 volts or more.

Accident additionally fabricate meters that can catch the pinnacle level of short drifters for example - The Fluke-87-5, Fluke-287 and Fluke-289 models.

GENUINE RMS MULTIMETERS

The voltage gracefully to your house is AC, and voltage and current differ in extremity after some time. The waveform is sinusoidal as in the graph underneath and the alter of course of current is known as the recurrence and estimated in Hertz (Hz). This recurrence can be 50 or 60 Hz, contingent upon which nation you live in. The RMS voltage of an AC waveform is the viable voltage and like the normal voltage. In the event that the pinnacle voltage is Vpeak, at that point the RMS voltage for a sinusoidal voltage is Vpeak/√2 (approx 0.707 occasions the pinnacle voltage). The force in a circuit is the RMS voltage duplicated by the RMS current streaming in a heap. The voltage regularly imprinted on apparatuses is the RMS voltage despite the fact that this isn't typically expressed.

An essential multimeter will show RMS voltages for sinusoidal voltage waveforms. The flexibly to our homes is sinusoidal so this isn't an issue. Be that as it may if a voltage is non sinusoidal, for example a square or triangular wave, at that point the meter won't demonstrate the genuine RMS voltage. Genuine RMS meters anyway are intended to effectively demonstrate RMS values for every single molded waveform.

The AC Supply Feeding Our Homes is a Sine Wave

Estimating Voltages Remotely and Logging Readings

In the event that you have to gauge voltages and log them after some time, you can utilize a datalogging multimeter. An item, for example, the Fluke 289 True-RMS datalogging multimeter can record 15,000 readings. Another component of this meter is that it very well may be arrangement with a remote connector to speak with an Android cell phone, permitting readings to be seen remotely, while the meter is found somewhere else.

CHAPTER SEVEN

ANSWERS TO FREQUENTLY ASKED QUESTIONS ABOUT THE MULTIMETERS

How Do You Check Voltage With a Multimeter?

Fitting the dark test into COM and the red test into the attachment checked Vω. Set the range to DC or AC volts and contact the test tips to the two focuses between which voltage should be estimated.

How Do You Check if a Wire is Live With a Multimeter?

For this current it's ideal to remain safe and utilize a non-contact volt analyzer or stage analyzer screwdriver. These will demonstrate if voltage is e.g > 100 volts. A multimeter can just gauge the voltage among live and unbiased or live and earth if these conductors/terminals are open, which may not generally be the situation.

How Do You Check Voltage Drop With a Multimeter?

Voltage drop happens over an obstruction or along a force link. So follow a similar methodology with respect to estimating voltage and measure voltage at the two purposes of interset and take away one structure the other to quantify voltage drop.

For what reason is Voltage Drop Important?

In the event that voltage drop is over the top, machines may not work appropriately. Link ought to be measured satisfactorily to limit voltage drop for the current it needs to convey and the separation over which current ventures.

GRAPHING MULTIMETERS: THE ULTIMATE TIME MACHINE

Getting information and plotting it to a screen after some time is perhaps the surest methods of nailing those quick happening intermittents. Also, that is the place a decent charting meter truly sparkles!

Have you at any point driven around with a planning light in one hand and a fuel pressure check taped to the windshield, wanting to decide whether that irregular slowing down issue you're facing is fuel-or start related? I have, and seldom was I happy with the outcomes. The planning light terminating was flighty to the point that when the slow down happened, I was never fully sure if the start was to blame or if the planning light trigger was simply turning up missing.

Enter the charting multimeter, a little thingamabob that spares long periods of time and disappointment and can be one of the most impressive demonstrative devices in your stockpile.

This is what I mean. Fig. 1 above is an account of my Eagle Premier while observing rpm and fuel siphon voltage. Note that when the slow down happened, rpm dropped like a stone. We realize that when a motor slows down as a result of a fuel issue, it commonly takes a couple of moments to lose rpm. A quick drop, then again, for the most part shows loss of start.

This sign was observed from the start loop optional wire, so we realize something happened to the start some place in the circuit preceding this point. It could be power, trigger, an open, a short or any number of different issues, however in any event it's limited to the start framework.

Note that this account was caught utilizing a product program called Waveform Manager Pro. Waveform Manager Pro interfaces a PC with test gear (for this situation, the OTC Perception charting meter), giving preferred goal over the presentation on the instrument alone. Both the meter screen catch and the PC recording are indicated next to each other. Notice that fuel siphon juice was lost right around 6 seconds after the start kicked the bucket.

There's no advanced science in Fig. 2 on the following page. Here we're looking at a starter framework by diagramming battery voltage (top outline) and starter current draw (base). Simply attach the meter, bounce in the vehicle, floor the quickening agent to enter clear flood mode (to forestall beginning) and go. It's a quick, one-man occupation, and there's no compelling reason to interface jumper prompts wrench the motor.

One note: You do need to disregard the high and low pinnacles and focus on the normal qualities. The top outline shows that the pinnacle voltage dropped to the second vertical division, or 7.5 volts, while the normal voltage drop when turning was at the third division, or 10.25 volts. The pinnacle amperage was around four divisions, or 200mV. The inductive test was set up to a 1mV-per-amp scale. So the pinnacle current draw was

200 amps, with the normal around 100 amps. No issue here.

The account in Fig. 3 on page 46 is of a Snap-on Vantage charting the obstruction of the cooling fan engine on my Eagle Premier. Gradually turning the fan past every commutator bar shows a predictable perusing from bar to bar. This demonstrates the engine windings are not shorted or open. This is a simple, quick test to perform and will take out those annoying irregular cooling fan engines that won't please because of a terrible winding. Of course, this test could have been performed with a current test and a degree, however here and there you utilize what's within reach. I utilized this model on the grounds that most charting meter clients once in a while utilize more than voltage settings. Grow the potential outcomes. Record each estimation you take, including obstruction.

Think about Some of the Possibilities

A diagramming meter is one of the most flexible bits of test hardware you can possess. Here are a portion of the utilizations I've found for

these little marvel boxes; I'm certain you can consider others, well.

• I utilized an AC voltage record to locate a discontinuous force drop on my shop's electric assistance. The service organization tech that showed up took one gander at my old assistance board and was certain that was the issue. At that point I demonstrated him the account from the voltage drop. He remarked, "Wish we had decent meters that way!" That drop was slaughtering my air blower, so I was happy to get it fixed...and for nothing, for sure!

• The coke machine in the shop nearby ran constantly, utilizing a lot of intensity. The retailer needed another machine, yet the provider demanded there was not all that much. In the wake of observing the over the top run time on the blower and faxing a duplicate to the provider, another machine was conveyed immediately.

• I regularly utilize my CRT current test and the Snap-on Vantage to discover discontinuous current draw issues. Moving around and squirming wires and additionally shaking the vehicle has nailed more than one damaged trunk or glovebox light switch.

• Monitoring cooling fan activity versus the ECT sensor perusing has spared me from a misdiagnosis time and again. Simply set up the diagramming meter, let it run, at that point keep an eye on it later.

• Recording fuel siphon pressures after some time utilizing a diagramming meter and weight transducer is a certain fire approach to catch those siphons that need a touch of heating up before they fire misbehaving. Or on the other hand record the fuel siphon current with a test. Even better, record both.

• The folks on iATN have done marvels diagnosing extreme a/c issues by recording high-side and low-side readings with pressure transducers snared to their diagramming meters.

•Compare injector beat width patterns versus O2 sensor, ECT, TPS, MAP or MAF readings. Obvious targets here! Fig. 4 at left is a chronicle of pinnacle terminating kV on my Eagle Premier during a snap choke test. This was taken with the new Snap-on Vantage kV Module. While a diagramming meter may not supplant your admired start scope, it sure can prove to be useful for some fast indicative checks. This clever little module snaps directly on the rear of the Vantage.

For me, the granddaddy of all charting meters is the custom meters capacity of Edge Diagnostic Systems' Simu-Tech. The capacity to make and control up to 20 meters on the double utilizing numerous sources of info can't be beat. Fig. 5 above is an ideal model. This is a playback of a Ford truck with a criticism carburetor framework. FCS is the fuel control solenoid. Notice the total powerlessness of the input framework to keep up fuel control out of gear after a decel occasion.

Highlights to Consider

Diagramming meters offer a large group of helpful highlights. It's dependent upon you to conclude which are generally significant.

When looking for such a meter, here are a couple of interesting points:

•Screen goal, test speed and record length.

• Input impedance.

•PC screen catch and interface capacities.

•Battery life, outer force gracefully alternative.

• Add-on embellishments (weight, vacuum and temperature transducers)/expandability.

• AC voltage capacity. Additionally note that some charting meters have databases of vehiclespecific data, for example, auto arrangements, sensor association focuses and operational portrayals. This truly grows the intensity of these instruments and ought to consistently be viewed as when settling on a buy choice. Make a few inquiries and look at the young men on iATN to settle on an educated choice.

For certain undertakings, utilizing a charting multimeter may appear chasing butterflies with a shotgun. In any case, when that large, terrible bear appears, you'll be happy you came prepared!

Picking The Right Digital Multimeter

Picking the privilege advanced multimeter (DMM) requires contemplating what you will utilize it for. Assess your essential estimation needs and employment necessities and afterward investigate exceptional highlights/capacities incorporated with numerous DMMs. Consider whether you need it for essential estimations or for further developed investigating choices.

Viewpoints to consider

Goal and exactness. Goal of a DMM is communicated in the quantity of digits the unit can show. For instance, a 4½ DMM has four full digits that show esteems from 0 to 9 and the partial digit, which is the most huge digit in the presentation. Fragmentary digit is either 0 or 1. Such a meter shows positive or negative qualities from 0 to 19,999.

Precision of a DMM is not the same as show goal. It is the most extreme permissible constraint of mistake in readings. All DMM producers express exactness determinations as ±(% of reading+number of least huge digit).

Working condition.

Estimations types. Practically all DMMs make voltage, current and opposition estimations. Most make precise AC estimations with sinusoidal signs, yet when such signals are not basic sine waves, their exactness endures. On the off chance that you have to make AC estimations of signs that have a ton of consonant mutilation, you might need to buy a DMM that makes genuine rms AC estimations. These cost all the more however will make progressively exact estimations.

Numerous DMMs are intended to take care of complex issues in gadgets, plant robotization, power appropriation and electromechanical gear. With the capacity to log information and survey it graphically onscreen, you can take care of issues quicker and help limit personal time. While choosing a computerized multimeter (DMM) for information logging, significant particulars incorporate the speed at which the DMM can make estimations and the measure of memory it requires.

To work securely approach electrical boards while wearing less close to home defensive hardware, or to maintain a strategic distance from the danger of being impeded while pursuing interrelated occasions, one needs to adjust to remote DMM.

Genuine rms versus normal reacting DMMs. Fundamentally there are two kinds of DMMs: normal reacting and genuine rms. The last is a reliable and standard approach to quantify and look at dynamic signs of every kind imaginable, though the previous is adjusted for sine-wave inputs as it were.

Normal reacting meters ordinarily function admirably for direct loads (standard enlistment engines, opposition warming, radiant lights and that's just the beginning), yet on the off chance that non-straight loads (electronic/electric-release lighting, movable speed drive frameworks, etc) are available, blunders happen, which could make the readings lower than anticipated. In the event that you need to gauge non-straight loads, for example, those found at electronic controls, you ought to pick genuine rms.

DMM input impedance. It is particularly imperative to choose a DMM with high impedance for applications requiring estimation of touchy gadgets or control circuits to guarantee precision of the readings.

CHAPTER EIGHT

DIGITAL MULTIMETER USES AND FUNCTIONS

I needed to share a brisk post with respect to utilizing a computerized multimeter. This is planned to be a short prologue to how multimeters work and their fundamental capacities. Each circuit repairman understudy should realize the fundamental capacities and how to utilize a computerized multimeter.

This post is intended to give you a fundamental comprehension of what an advanced multimeter is and how to utilize one. Make a point to likewise investigate the different multimeter images that various brands use.

What Is A Digital Multimeter?

An advanced multimeter or DMM is one of the most helpful things of test hardware to analyze electrical or electronic issues. It's an estimating gadget that takes simple data and utilizations a simple to advanced converter to change over it into a computerized signal which peruses out on the presentation.

You can utilize an advanced multimeter for estimating the accompanying:

- AC voltage
- DC voltage
- Current (both AC current and DC current)
- Resistance
- Temperature
- Frequency
- Capacitance

Obviously, this all relies upon the sort of advanced multimeter you have. At any rate, you ought to have the option to gauge voltage, obstruction, and current with any computerized multimeter.

The most effective method to Use A Digital Multimeter To Test Voltage

Testing a circuit (or wellspring of intensity) for voltage is the most well-known utilization of an advanced multimeter. Here's the way you would test various circuits for voltage.

DC or Direct Current Circuits

1. Install your test leads into the best possible terminals on the meter. Dark lead into the normal or COM port and the red lead into the port marked Vω.

2. Turn on your multimeter and spot the selector change to the V with the strong and ran line (VDC).

3. Now associate the test prompts the circuit: dark to the negative likely point (which is the circuit ground) and red to a positive expected point. Credit:

You should now observe a perusing on the showcase. The presentation is currently indicating the voltage of the source (battery) or circuit. On the off chance that you see a − image before the perusing, at that point switch your leads (this implies the extremity is in reverse). In the event that you don't see a perusing, at that point check to ensure you've chosen the right setting and the leads and the leads are associated accurately. Still no perusing? The source is either dead or the circuit is off. In the event that it's still not perusing… at that point your multimeter is busted (ensure you get extraordinary compared to other multimeters available... there is a great deal of garbage out there).

Air conditioning or Alternating Current Circuits

1. Install your test leads into the best possible terminals on the meter. Dark lead into the normal or COM port and the red lead into the port marked Vω.

2. Turn on your multimeter and spot the selector change to the V with the wave.

3. Since it's AC, the lead arrangement doesn't make a difference yet you despite everything need to comprehend what readings to expect on conductors or containers.

THE END

CPSIA information can be obtained
at www.ICGtesting.com
Printed in the USA
LVHW080204260822
726924LV00013B/1153